101 Apple Recipes

By Carole Eberly

101 Apple Recipes

Denali & Co., LLC. PO Box 1015, Traverse City, MI 49685

ISBN: 979-8-9852610-4-2

Hi!

Just a few notes before you grab your apple peeler and pie pan.

You'll notice the scarcity of vegetable recipes here. (I did include two so my mother wouldn't feel she utterly wasted her time on all those childhood lectures about such things.) However, I made up for it by including 16 pie recipes covering everything from the downright sinfully gooey to "Plain Old Apple Pie."

Use cooking apples for all recipes unless they say differently.

Most of all, have a great time with this book. I did.

Carole

Table of contents

A few Apple definitions I couldn't resist:

Apple Cation	What you fill out when interviewing for a job
Apple Rition	A ghost
Apple Itionist	A person opposed to slavery
Apple Ene	Eisenhower's city
Apple Latia	A geographical region in North America
Apple Oosa	A kind of horse

A Few Apple Facts To Dazzle Your Friends With:

- More than 7,500 varieties of apples are now being produced.
- Apples are not native to North America. The first trees were planted by the first European settlers.
- Johnny Appleseed's apples were pretty awful tasting because the good ones are grown from grafts, not seeds. (Sorry, Johnny).
- Our stone age friends knew how to preserve apples by freezing and sun drying them.
- American apple growers produce over 255 million bushels annually.
- Rub an apple slice on a wart to make it disappear. (If you believe that, I've got some real estate I want to sell you).

What Equals What

Use six to eight medium apples for a 9" pie.
Use two pounds of apples for three cups of applesauce.
Use one bushel of apples for 16 to 20 quarts of applesauce.
Two large, three medium, or four small apples make a pound.
Two-thirds of a pound makes 2 cups sliced apples.

(Got all that?)

Which apples to choose:

Variety	**Characteristics**	**Season**	**Uses**
McIntosh	Deep red, sometimes striped with green. Slightly tart.	Sept-Feb	Eating, baking, salads
Jonathan	Bright red, small to medium sweet-tart flavor.	Sept-Feb	Eating, baking, pies, salads
Red Delicious	Bright red, conical shaped with five point on bottom. Sweet.	Sept-April	Eating

Variety	Characteristics	Season	Uses
Golden Delicious	Yellow, conical shaped with five points on bottom. Juicy, sweet.	Oct-April	Eating
Northern Spy	Bright striped and mottled red on yellow. Large. Very juicy.	Oct-March	Eating, baking, excellent in pies
Baldwin Red	Mottled with bright red. Mildly tart. Juicy.	Nov-April	Eating, cooking
Ida Red	Bright, red, smooth. Juicy.	Oct-June	Eating, pies, salads, baking
Rome Beauty	Bright red with yellow. Slightly tart. Round. Stores well by bushels.	Nov-May	Baking, pies
Wealthy	Bright red splashes over green skin. Medium. Juicy.	Aug-Dec	Eating, pies
Winesap	Deep red tinged with yellow. Very juicy. Slightly tart.	Jan-May	Eating, pies

Snacks
Breakfast
Breads
Beverages

Apple Quiche

Everyone will think you worked really hard! Ha Ha!

1 9" unbaked pie shell
5 large apples
1 cup cheddar cheese, shredded
2 eggs
1 cup evaporated milk
1/2 teaspoon salt
Dash nutmeg

Peel, core, and slice apples into thin pieces. Arrange in pie shell. Sprinkle with cheese. Mix eggs, milk, salt, and nutmeg together in small bowl. Pour over cheese. Bake at 350 degrees for 60 to 70 minutes. Serve hot or cold.

Apple Pizza

11-oz. package pie crust mix
5 medium apples
1/2 cup flour
1/2 cup brown sugar
1/2 teaspoon cinnamon
1/4 cup butter
1/2 cup chopped walnuts

Make up dough for 2-crust pie and roll out into one 14" round. Place on ungreased 12" pan; turn under edges and flute. Core and slice apples on pie shell.

Stir remaining ingredients, except nuts, and crumble over apples. Sprinkle walnuts over pie and bake at 450 degrees for 25 minutes. Serve with ice cream or cheddar cheese.

Apple Butter and Cheddar Waffles

Frozen waffles
Apple butter
Cheddar cheese, shredded

Toast waffles. Spread each with 1-1/2 tablespoon apple butter and top with 1 tablespoon cheddar cheese. Place in broiler until cheese melts.

Apple and Cheese Snacks

Unpeeled apple rings
Cream cheese
Chopped walnuts

Spread rings with softened cream cheese. Sprinkle with walnuts.

Double Apple Sandwiches

Grated apples
Cream cheese
Pitted dates
Chopped walnuts
Apple bread

Mix apples, cream cheese, dates, and walnuts. Spread on slices of Apple bread.

Peanut Butter and Apple Sandwiches

Unpeeled apple rings
Peanut butter

Spread slices of bread with peanut butter. Layer on apple rings.

Apple & Cheese Canapés

5 medium apples
16 slices white bread
6-oz. cream cheese
1 5-oz. jar cheddar cheese spread
1 5-oz. jar olive-pimento spread
Lemon juice

Core apples and cut into 16 thin rings. Cut rounds out of bread same size as apples. Spread rounds with cream cheese. Place an apple slice on each. Brush with lemon juice. Mix cheese spreads together and fill centers of apple slices with spoonfuls of mixture.

Maple Apple Granola

3 cups oats
1-1/2 cups wheat germ
1/2 cup bran
1 cup dried apples
1/2 cup chopped walnuts
1/2 cup chopped almonds
2 cups powdered milk
1/4 cup sesame seeds
1/4 cup unsweetened coconut
1/4 cup salad oil
1/4 cup melted butter
1-1/2 cups maple syrup

Makes 8 - 1/2 cups

Mix ingredients and bake at 225 degrees for 1-1/2 to 2 hours, stirring occasionally.

Birchermuesli

Easier to make than pronounce.

1 cup oats
1 cup chopped almonds
2 cups yogurt
1/2 cup orange juice
2 lemons, juice and rind
4 tablespoons evaporated milk
4 grated apples
Brown sugar

Serves 4-6

Mix ingredients and sweeten to taste with brown sugar. Add one or more sliced fruits that are in season such a strawberries, bananas, or peaches.

Apple Fritters

- 1-1/2 cups flour
- 1 tablespoon sugar
- 2 teaspoons baking powder
- 1/2 teaspoon salt
- 3 eggs
- 2/3 cup milk
- 2 tablespoons salad oil
- 3 cups diced apples

Sift dry ingredients. Beat eggs, milk, and salad oil. Add dry ingredients and mix until just moistened. Stir in apples. Drop from tablespoon into hot (375 degrees) shortening. Fry till puffy and brown, 3 to 4 minutes, turning once. Drain on paper towels. Sprinkle with powdered sugar.

Apple Muffins

- 2 cups flour
- 1/4 cup sugar
- 1 tablespoon baking powder
- 1 teaspoon salt
- 1 teaspoon cinnamon
- 1 apple, diced
- 1/2 teaspoon nutmeg
- 1 egg
- 3/4 cup milk
- 1/4 cup salad oil

Makes 12

Mix dry ingredients with apple. Beat wet ingredients and stir in dry ones, stirring just until moistened. Fill muffin papers 2/3 full. Bake at 400 degrees for 20 to 25 minutes.

Apple Cornbread

1 cup cornmeal
1 cup flour
1/3 cup sugar
1 tablespoon baking powder
1 teaspoon salt
1 egg
1 cup milk
1/4 cup salad oil
1 apple, diced

Mix dry ingredients. Beat wet ingredients, stirring in dry ones. Add apple. Pour into well-greased 9" square pan. Bake at 425 degrees for 20 to 25 minutes.

Apple Pancakes

1 cup flour
1/2 teaspoon salt
1 teaspoon baking powder
1 egg
1 cup milk
1 tablespoon salad oil
1 apple, diced

Mix dry ingredients. Beat wet ingredients and stir in dry ones. Mix in apple. Bake on hot greased griddle.

Crunchy Apple Coffee Cake

1/4 cup flour
1/4 cup brown sugar
1/2 teaspoon cinnamon
1/4 teaspoon nutmeg
2 tablespoons butter
1-3/4 cups flour
1/2 cup sugar
1 tablespoon baking powder
1/2 teaspoon salt
1/2 cup butter
3/4 cup milk
1 egg, slightly beaten
2 apples, (peeled and sliced thin)

Mix flour, brown sugar, cinnamon, and nutmeg in small bowl. Cut in 2 tablespoons butter until mixture is crumbly. Set aside. Combine flour, sugar, baking powder, and salt in bowl. Cut in 1/2 cup butter. Add milk and egg, stirring until moistened. Spread batter in greased 9" square pan. Arrange apple slices on top. Sprinkle with topping. Bake at 400 degrees for 30 to 35 minutes.

Quick Apple Bread

1/2 cup shortening
1 cup sugar
1-1/2 tablespoons sour milk
2 eggs
1 teaspoon vanilla
1 teaspoon baking soda
1 teaspoon salt
2 cups flour
1 cup cider
1-1/2 cups apples, chopped
1/2 cup walnuts, chopped
2 teaspoons sugar
1/2 teaspoon cinnamon

Cream shortening and sugar. Add sour milk, eggs, and vanilla, blending well. Alternate adding dry ingredients with cider. Stir in apples and walnuts.

Pour into a greased loaf pan. Sprinkle with mixture of sugar and cinnamon. Bake at 350 degrees for 1 hour, 15 minutes.

Applejack

Hard cider

Put hard cider outside to freeze in winter. Whatever doesn't freeze is applejack.

Cider Punch

1 cup sugar
1/4 cup lemon juice
2-1/2 cups cider
1 cup orange juice
2-1/2 cups ginger ale

Heat sugar and lemon juice in saucepan until sugar dissolves. Add to cider and orange juice. Chill. Add ginger ale before serving.

Hot Mulled Cider

The best!

- 1 gallon cider
- 1 cup brown sugar
- 1/2 cup butter
- 1 stick cinnamon, broken
- 2 teaspoons whole cloves

Bring cider to a boil. Add remaining ingredients and simmer over low heat, stirring until sugar dissolves.

Cranberry Cider Punch

- 1 cup sugar
- 1/2 cup water
- 1 quart cider
- 2 cups cranberry juice
- 1/4 cup lemon juice
- 1 quart ginger ale

Boil sugar and water in saucepan. Add to the cider and juices. Chill. Before serving, add the ginger ale.

Cognac Cider Punch

1 cup orange juice
1 cup lemon juice
1 cup pineapple juice
1/4 cup maraschino cherry juice
2 quarts cider
1 cup cognac

Mix all ingredients together and serve over a block of ice.

Apple Refresher

1 quart apple cider
1 pint vanilla ice cream
1/2 teaspoon cinnamon

Mix ingredients together in blender until frothy. Serve immediately.

Serve 2 to 3

How many Apple clichés can I get into one paragraph!

“

Why, he’s as American as apple pie. He’s the apple of my eye, even though he’s rotten to the (apple) core. You know what they say, ‘One bad apple spoils the barrel.’ He’s a bad apple for sure. He’s an apple polisher and wants everything in apple pie order. When we took a trip to the Big Apple, he was always upsetting the apple cart. When I drank the water and came down with the green-apple-quick-step, he kept saying, ‘An apple a day, keeps the doctor away.’ Still, he believes in apple pie and motherhood.

”

Whew!

(I couldn’t figure out where to put in Adam’s apple.)

Preserves
Jams
Sauces and Salads

Freezing Apple Slices

Enough for three 9” pies

Wash firm, mature apples. Peel, core, and slice. Soak in a solution of 2 tablespoons salt per gallon water for 15 minutes. Rinse and drain. Dry with paper towels. Cover all surfaces with sugar and pack in freezer containers.

For a frozen apple pie, filling try this: Peel, core, and slice 18 large apples. Put in a pot with 6 tablespoons butter, 3 cups sugar, 3 teaspoons cinnamon, and 1 teaspoon nutmeg. Cook over very low heat 15 minutes, stirring occasionally. Cool filling. Pour into three freezer containers.

I would have made this a separate recipe, but then I’d have to change the title of this book.

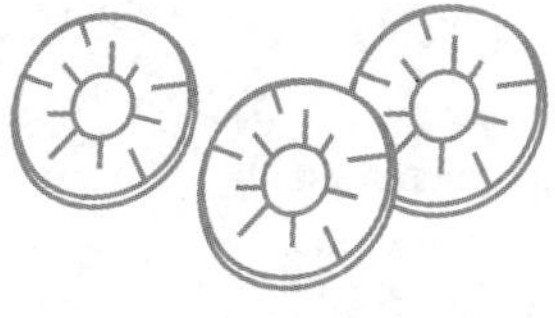

Sun Dried Apples

Wash firm, mature apples - preferably Northern Spies or Baldwins. Core and slice into rings. Brush well with a solution of 2,000 mg of a ascorbic acid (Vitamin C tablets will work) to a quart of water. Place on a clean cloth and cover with a canopy of cheesecloth. Place in a spot that receives sun all day. Drying time depends on temperature, sun, and humidity– which ideally, should be low. Test fruit by tearing it apart. It should have a flexible "kid glove" texture. When done "pasteurize" it by putting it on baking sheets in a very low –175 degrees– oven. This will get rid of any little bad things that want to share your apple rings.

If you don't see the sun much where you live because of pollution, after coating the dry apple slices with ascorbic acid, put them in the oven at a very low setting– not more than 175 degrees. Apple slices need air to circulate around them so put them on cake cooling racks to make a mesh screen. Keep the door ajar while baking. When they're through, 4 to 14 hours, process as above. Package in plastic bags.

Baked Apple Slices

6 medium apples
1/4 cup water
1/3 cup brown sugar
1 teaspoon cinnamon

Peel and core apples. Cut into rings and place in large saucepan. Pour on water, sprinkle with sugar and spices. Cover and simmer for 20 to 30 minutes.

Fried Apples

6 large apples
4 tablespoons bacon fat
1/2 cup sugar
1/4 teaspoon salt
1/4 cup water

Core and slice apples into thick pieces. Heat bacon fat in frying pan and drop apple slices in. Cover and cook until brown. Add sugar, salt, and water. Simmer until apples are tender

Batter Fried Apple Rings

1 cup flour
1-1/2 teaspoons baking powder
2 tablespoons sugar
1/2 teaspoon salt
3/4 cup milk
1 egg
4 large apples
Sugar and cinnamon

Sift dry ingredients. Add milk and egg, beating well. Peel and core apples. Slice into 1/4" rings. Dip rings in batter and fry in 1/2" hot shortening until golden brown. Drain on paper towels. Sprinkle with sugar and cinnamon mixture.

Fried Green Apples

4 tablespoons shortening
6 green apples
3/4 cup sugar
1/2 teaspoon cinnamon

Heat shortening in frying pan. Core and slice apples. Drop into skillet with sugar and cinnamon. Cook slowly, stirring occasionally until apples are browned.

Curried Apple Slices

3 cups apple slices
1/4 cup melted butter
1/4 cup sugar
1 tablespoon curry powder

Mix all ingredients together. Broil 10 to 12 minutes. Great with meats.

Cinnamon Apple Rings

1/4 cup sugar
1-1/2 cups water
1/2 cup red cinnamon candies
4 apples

Combine sugar, water and candies in frying pan, heating until the candies melt. Core apples and slice into rings. Simmer in syrup until transparent.

Spiced Crab Apples

Makes 6 pints

4 cups sugar
2 cups water
1 cup vinegar
2 sticks cinnamon, broken
2 teaspoons whole cloves
1-1/2 quarts crab apples

Heat sugar, water, vinegar, cinnamon, and cloves to boiling. Drop 1/6 of the apples into syrup and cook until tender. Pack in hot, sterile jar. Add hot syrup to 1/2" from top. Seal. Repeat with remaining apples.

Crab Apple Preserves

1 lb. crab apples
1 lb. sugar
1 cup water
1/2 lemon, slice thin

Peel crab apples. Boil sugar, water, and lemon peel. Add crab apples and cook until tender. Remove from heat and let stand overnight. Pack fruit into jars, pouring syrup on top. Seal.

Apple Relish

6 cups chopped apples
2 green peppers, chopped
1 cup onion, chopped
2 red peppers, chopped
1 teaspoon mustard
1/2 cup vinegar
1 cup water
1-1/2 cups sugar

Mix all ingredients together in pot. Cook slowly for 1/2 hour. Pack in hot, sterile jars and seal.

Old-Fashioned Apple Butter

5-lb. tart apples
3 cups apple cider
4 cups sugar
3 teaspoons cinnamon
1 teaspoon cloves

Makes about ter 6-oz. jar

Peel, quarter, and core apples. Bring apples and cider to a boil in heavy saucepan. Reduce heat and simmer, covered, for 20 to 25 minutes. Purée apple mixture in blender.

Add sugar and spices, blending well. Pour purée into a shallow baking dish and bake at 300 degrees for 2 hours, or until thick enough to hold its shape. Pour into hot, sterile jars and seal.

Apple Jelly

Apples
Water
Sugar

Quarter and core apples. Simmer in saucepan, with just enough water to cover, until soft. Strain out juice. Add 1 cup sugar for each cup liquid. Simmer until mixture reaches 220 degrees. Pour into hot, sterile jars and seal. Four cups of juice makes about 4 6-ounce jars.

Apple Marmalade

6 lb. tart apples
3 lemons
1-oz. fresh ginger root
3-lb. sugar
1/4 cup water

Core apples and slice into thin pieces. Cut peel from lemons into thin strips. Grate ginger. Put apples, peel, ginger, sugar, and water into pot. Squeeze in juice from lemons and bring to a boil. Cook at low boil until syrup is thick. Stir often. Pour into hot, sterile jars and seal.

Instant Applesauce

Good!

1/4 cup cider
4 medium apples
1/4 cup sugar
1/8 teaspoon cinnamon

Peel, core, and slice apples into thick pieces. Put cider and apples into blender. When smooth, add sugar and cinnamon. Blend.

Honey Applesauce

Better!

12 medium apples
2-1/2 cups water
1/2 cup honey
1/2 teaspoon cinnamon
1/4 teaspoon nutmeg

Peel, core and slice apples into thin pieces. Cook with water in saucepan until soft. Stir in honey and spices, simmering over low heat until fairly thick.

Chunky Applesauce

Peel, core, and slice 6 apples into thick pieces. Cook with just enough water or cider to keep from scorching in a covered saucepan. Stir occasionally. Mash up apple pieces a little. Sweeten to taste with brown sugar. Add a dash or two of cinnamon and nutmeg.

Best!

Smooth Applesauce

Follow the above recipe, but grind apples through a food mill at the end of cooking. Add sugar and spices to taste.

Liberate yourself from that yucky, canned stuff in the stores. Once you make your own applesauce, you'll never buy another can.

Waldorf Salad

2 cups diced apples
1 cup diced celery
1/2 cup chopped walnuts
1/4 cup mayonnaise
1/2 cup whipped cream

Mix apples, celery, and walnuts with mayonnaise and whipped cream.

Apple Slaw

1 Spanish onion
2 apples
4 cups shredded red cabbage
1/4 cup dill pickle spears
French Dressing

Slice onion into rings. Core apples and slice thinly into rings. Combine onion, apple, cabbage, and pickle. Moisten with French dressing.

Golden Delicious Salad

1 6-oz. package lemon gelatin
1-1/2 cups boiling water
2 cups orange juice
3/4 pint cottage cheese
1/2 cup chopped nuts
2 cups diced golden delicious apples

Dissolve gelatin in water. Add orange juice. Pour 1" layer into 2 quart mold. Chill until firm. Chill remaining gelatin until syrupy. Add remain ingredients and pour over firm gelatin. Chill until set.

Red Apple Salad

4 apples
1-1/2 cups sugar
1/2 cup red cinnamon candies
3 cups water
1/2 cup cottage cheese
1/4 cup chopped green pepper

Pare and core apples. Heat sugar, candies, and water until candies dissolve. Cook apples slowly in covered pan in this syrup. Turn apples occasionally. Drain and chill. Stuff apples with cottage cheese and green pepper.

Bavarian Apple Cabbage

1 medium apple
1 medium head red cabbage
3 tablespoons butter
1 bay leaf
1/4 cup white wine
2 tablespoons vinegar
1 teaspoon sugar
Salt and pepper

Serves 6

Peel, core, and slice apple. Shred cabbage and add to butter with apples, mixing well. Cover and cook on low heat for 10 minutes.

Add remaining ingredients. Cover and cook for 45 minutes, stirring occasionally.

Vegetables Stuffing and Main Dishes

Scalloped Sweet Potatoes and Apples

3 medium sweet potatoes
1/2 cup butter
3/4 cup brown sugar
1/2 cup apple cider
1/2 teaspoon mace
2 medium apples, pared and sliced

Serves 6

Boil potatoes in water for 25 minutes. Drain and let cool. Peel and cut into 1/4" slices. Melt butter in small pan. Add brown sugar, cider, and mace. Bring to a boil, stirring until sugar dissolves. Simmer, uncovered, 10 minutes.

In baking dish, layer potato and apple slices. Spoon half of syrup over potato mixture. Bake, covered at 400 degrees for 25 minutes. Remove cover and bake 25 more minutes, basting with remaining syrup.

Apple Raisin Stuffing

1/2 cup minced onion
2 cups diced apples
4 cups soft bread crumbs
1/2 cup raisins
1 teaspoon salt
2 teaspoons poultry seasoning
1/2 cup melted butter

Combine all ingredients, mixing well. Makes enough to stuff 6- to 7-lb. capon or 8- to 10-lb. turkey.

Sausage Apple Stuffing

1/2 cup sausage
1/2 cup diced apples
2 tablespoons chopped onion
1 cup soft bread crumbs
Salt and pepper
Hot water

Fry sausage, apple and onion until cooked. Add bread cubes, salt and pepper to taste and enough hot water to make a good stuffing. Enough for a 2- to 3-lb. chicken.

Apple Stuffing

2 tablespoons butter
4 tablespoons chopped onion
4 diced apples
4 cups bread cubes
2 eggs
1/2 teaspoon nutmeg
Salt and pepper
2 tablespoons parsley, chopped
Water

Cook onions and apples in butter until tender. Add bread cubes, eggs, nutmeg, salt and pepper to taste. Add parsley. Mix well. Add enough water to make a good stuffing. Enough for a 6- to 8-lb. turkey.

Apple Stuffed Pork Chops

6 pork chops, with pockets for stuffing.
2 cups stuffing
Cooking oil
1/4 cup water

Put 1/3 cup stuffing (Apple, Sausage Apple, or Raisin Apple) in pocket of each chop. Secure with wooden picks. Brown chops on both sides in oil. Add 1/4 cup water and simmer for 45 to 50 minutes, or until tender.

Sherried Chicken and Apples

3 large apples
1 tablespoon lemon juice
1/4 cup sherry
1/4 cup brown sugar
Flour
1/2 cup butter
Sugar and cinnamon
1/2 cup flour, divided
1 teaspoon paprika
1 teaspoon salt
1/4 teaspoon pepper
1/4 cup salad oil
1/2 cup apple cider
1/4 cup sherry
2 cups light cream
1 frying chicken, cut up

Serves 4

Core apples and cut into sixths. Mix lemon juice, sherry, and brown sugar. Pour over apple wedges and let stand 1 hour. Drain apples and dip in flour. Fry slowly in butter until golden and tender. Sprinkle with sugar and cinnamon. Serve with chicken. Combine 1/4 cup flour, paprika, salt and pepper.

Coat chicken pieces. Brown in salad oil. Add cider, cover and simmer until tender, about 30 to 40 minutes. Add sherry. Cook 2 minutes. Remove chicken to platter. Add remaining flour to pan and mix with drippings. Add cream. Stir over low heat until smooth and thick. Spoon over chicken.

Apple Meat Loaf

2-1/2 lb. ground beef
1-1/2 cups bread crumbs
2 cups diced apples
1 egg
1/4 cup minced onion
2 tablespoons mustard
2 tablespoons Worcestershire sauce
1/3 cup ketchup

Serves 6 to 8

Mix all ingredients thoroughly. Bake in a greased loaf pan at 350 degrees for 1-1/4 hours.

Applesauce Meat Balls

Serves 6 to 8

Follow the Apple Meat Loaf recipe, substituting 1-1/2 cups applesauce for the diced apples. Shape into 2" balls and place in baking dish. Pour 1/3 cup ketchup and 1/4 cup water over top and bake at 350 degrees for 45 minutes.

Apple and Ham Casserole

- 3-1/2 cups ground cooked ham
- 1 cup soft breadcrumbs
- 1 egg
- 1/2 cup milk
- 1 tablespoon mustard
- 1/4 teaspoon ground cloves
- 1 tablespoon minced onion
- 3 apples
- 1/3 cup brown sugar
- 2 tablespoons butter

Serves 4 to 6

Mix together ham, bread, egg, milk, mustard, cloves and onion. Put in baking dish. Peel, core, and slice apples; arrange apples on top.

Sprinkle with brown sugar and dot with butter. Bake at 375 degrees for 40 minutes.

Baked Apples and Sausage

Apples
Brown sugar
Sausage meat

How simple can you get?

Core apples and peel 2" strips from tops. Sprinkle centers with brown sugar. Stuff with sausage meat and bake at 350 degrees for 1 hour.

Apple Pork Balls and Sauerkraut

1 egg
1-lb. ground pork
1/4 teaspoon nutmeg
Dash cinnamon
1 medium apple
1/4 cup corn meal
2 tablespoons butter
2 cups sauerkraut
1 cup sour cream

Serves 4

Beat egg. Mix in pork, nutmeg, and cinnamon. Peel, core, and grate apple. Mix ingredients together and form 2" balls. Roll in corn meal. Heat butter in frying pan and add meat balls. Brown well. Reduce heat to low, cover and cook 30 minutes.

Heat sauerkraut in saucepan. Mix sour cream in with pork balls. Heat, but don't boil. Drain sauerkraut and place on serving dish. Spoon pork balls and sauce over top.

Meat & Apples Rings

1 lb. ground pork
1/2 lb. ground beef
2 tablespoons chopped onion
1 egg
1/2 teaspoon salt
1 apple

Mix pork, beef, onion, egg, and salt. Shape into 12 patties, Peel, core, and cut apple into six rings. Place 1 ring on each patty. Cover with another patty, pressing to seal edges. Bake at 350 degrees for 45 minutes.

Desserts and Gooey Things

Apple Pandowdy

Super!

4 large apples
1 cup brown sugar
1/2 teaspoon cinnamon
3 tablespoons butter
1-1/2 cups flour
2 teaspoons baking powder
2 tablespoons sugar
1 teaspoon salt
6 tablespoons butter
2/3 cup milk

Serves 6

Peel, core, and slice apples in bottom of greased 8" x 8" baking pan. Sprinkle with brown sugar and cinnamon. Dot with butter. Sift flour, baking powder, sugar, and salt. Cut in butter. Add milk to make soft dough. Take out small pieces of dough and press between the palms of your hands. Lay on top of apples, pressing dough together to form 1 big piece. (It sounds dumb, but it works.) Bake at 350 degrees for 45 minutes. Cut into squares and turn upside-down to serve. Serve with nutmeg -flavored cream.

Applesauce Brownies

This is actually more of a cake than a brownie, but it is brown and I had to call it something.

1-1/4 cups sugar
1/2 cup butter
2 cups flour
1-1/4 teaspoons baking soda
2-oz. melted bitter chocolate
1/2 teaspoon cinnamon
2 cups applesauce
2 eggs
1/2 cup chopped walnuts
6-oz. package chocolate chips

Beat sugar and butter together. Sift dry ingredients. Add to sugar and butter with chocolate, applesauce, and eggs. Beat well. Pour into greased 9" x 13" pan. Sprinkle with walnuts and chocolate chips. Bake at 350 degrees for 30 minutes.

Apple Bars

1 cup flour
1/2 teaspoon salt
1/2 teaspoon baking soda
1 teaspoon cinnamon
1-1/2 cups uncooked oats
2/3 cup brown sugar
1/2 cup butter
1 egg
1 teaspoon vanilla
1/4 cup chopped walnuts
2 cups thinly sliced pared apples
Powdered sugar

Sift flour, salt, baking soda, and cinnamon. Beat in oats, sugar, butter, egg, and vanilla. Press half of dough in bottom of greased 9" x 9" square baking dish. Sprinkle walnuts over dough. Arrange apple slices over nuts. Roll remaining dough between waxed paper to form 9" square. Place dough over filling and press lightly around edges. Bake at 350 degrees for 25 to 30 minutes. When cool, sprinkle with powdered sugar. Cut into bars.

Apple Butter Bars

Follow the recipe above, but substitute 1-1/2 cups apple butter for the apple slices.

Apple Dumplings

A lot of work,
but they sure taste good
on a cold, dull January day.

2 cups brown sugar
2 cups water
1/2 teaspoon cinnamon
1/2 cup butter
2 cups flour
2 teaspoons baking powder
1 teaspoon salt
2/3 cup shortening
1/2 cup milk
6 medium apples, peeled and cored
Sugar, cinnamon, nutmeg
Butter

Bring brown sugar, water, 1/2 teaspoon cinnamon, and 1/2 cup butter to a boil. Remove from heat. Sift flour, baking powder, and salt. Cut in shortening. Stir in milk and roll dough out 1/4" thick on lightly floured surface. Cut into 6" squares.

Put an apple on each square and sprinkle with sugar, cinnamon, and nutmeg. Dot with butter. Moisten edges of squares and fold corners to center, pinching edges together. Place on ungreased baking dish. Pour syrup over dumplings and bake at 375 degrees for 35 minutes. Serve with cream.

Apple Puffs

1 cup water
1 cup sugar
1/2 teaspoon vanilla
1-1/2 cups flour
2 teaspoons baking powder
1/4 teaspoon salt
1/2 cup butter
3/4 cup milk
2 tablespoons soft butter
2 tablespoons sugar
1 teaspoon cinnamon
5 apples, peeled and cored

Simmer sugar and water 5 minutes. Remove from heat and add vanilla. Sift flour, baking powder, and salt. Cut in 1/2 cup butter. Stir in milk for a soft dough. In a separate bowl, combine sugar, 2 tablespoons butter, and cinnamon. Slice apples into a greased 8" baking dish.

Pour syrup over apples. Drop dough by spoonsful over apples making 8 mounds. With a spoon, make a depression in each puff. Put some spice mixture into each. Bake at 425 degrees for 30 to 35 minutes until puffs are golden brown. Serve with cream.

Puffy Apple Pancake

3 eggs, separated.
3 tablespoons sugar
1/2 cup milk
1/4 cup flour
1/2 teaspoon salt
1/4 teaspoon baking powder
2 tablespoons butter
1 cup diced apple
1 teaspoon cinnamon
2 tablespoons sugar

Serve with Apple Cinnamon Syrup or Apple Walnut Syrup.

Serves 2

Beat egg whites until foamy; gradually add sugar and beat until stiff. Beat egg yolks, milk, flour, salt, and baking powder. Stir in apple and fold in egg whites. Melt butter in oven-proof frying pan.

Spoon mixture into hot pan. Sprinkle on cinnamon and sugar. Bake at 400 degrees for 5 minutes. Reduce heat to 375 degrees for 10 minutes.

Apple Walnut Syrup

3 tablespoons butter
1/4 cup chopped walnuts
2 cups apples, sliced and peeled
1/2 teaspoon cinnamon
Dash salt
1 cup maple syrup

Serve on waffles, ice cream, pancakes, custard, etc.

Melt butter. Add nuts to brown slightly. Remove nuts and add remaining ingredients. Cover and simmer about 10 minutes. Remove cover and simmer 3 more minutes. Remove from heat and add nuts.

Apple Cinnamon Syrup

2 tablespoons cornstarch
3/4 cup cold water
1-1/2 cups sugar
2 tablespoons lemon juice
3-1/2 cups peeled and sliced apples
1/2 teaspoon cinnamon

Mix cornstarch and water in saucepan until smooth. Stir in remaining ingredients and bring to a boil, stirring constantly. Cook until thick.

Apple Cobbler

1/2 cup butter
1 cup flour
1/4 teaspoon salt
1 cup sugar
2 teaspoons baking powder
1/2 teaspoon cinnamon
1/2 cup milk
4 apples

Melt butter an 8" square baking dish. Mix dry ingredients together. Add milk and beat until smooth. Peel, core, and slice apples. Pour batter into baking dish. Top with apple slices. Bake at 350 degrees for 45 minutes.

Apple Brown Betty

1/3 cup sugar
1/2 teaspoon cinnamon
1/4 teaspoon salt
2 cups fine dry crumbs
4 apples, pared and diced
3 tablespoons melted butter

Mix sugar, cinnamon, and salt. Put layer of crumbs in greased baking dish. Cover with layer of apples. Sprinkle with sugar mixture. Continue until all ingredients are used, having crumbs on top. Pour butter on top. Cover dish. Bake at 375 degrees for 40 minutes. Uncover last 10 minutes.

Candied Apple Halves

6 apples (Jonathan)
1 cup sugar
1 cup water
1/2 cup cinnamon candies

Halve and core apples. Place cut side up in glass baking dish and pour sugar over them. Add water. Fill centers of apples with candies. Bake at 350 degrees until candies have melted. Turn apples and continue baking until apples are transparent, about 1 hour.

Apple Crisp

8 medium apples
1/2 cup water
1 teaspoon cinnamon
1/2 teaspoon nutmeg
3/4 cup flour
1 cup brown sugar
1/2 cup butter

Peel, core, and slice apples thin. Place in greased casserole dish. Cover with water, cinnamon, and nutmeg. Blend other ingredients until crumbly. Sprinkle over apples and bake at 350 degrees for 40 to 50 minutes.

Red Candied Apples on a Stick

12 small apples
12 wooden sticks
3 cups sugar
3/4 cup corn syrup
1 cup water
4 drops oil of cloves
12 cinnamon candies

Insert stick in bottom of each apple. Heat sugar, corn syrup, and water in saucepan, stirring until sugar melts. And oil of cloves and candies. Cook, without stirring, until mixture forms a hard piece that cracks when dropped in cold water– 290 degrees on candy thermometer. Set pan over boiling water. Dip each apple into syrup. Let stand on buttered plate.

Caramel Apple

1-lb. caramels
2 tablespoons water
6 medium apples
6 wooden sticks

Guaranteed to cause cavities.

Melt caramels with water in double boiler. Insert stick in bottom of each apple. When caramels are melted and smooth, dip each apple into mixture, turning until well coated. (If desired, roll in chopped nuts.) Let dry on buttered plate or waxed paper.

Apple Leather

3 cups applesauce
1/2 cup sugar
1/2 teaspoon cinnamon
1/4 teaspoon nutmeg

Mix ingredients. Spread 1/4" thick in baking dish. Bake at 225 degrees about 3 hours. Cool and roll like a jelly roll. Cut into pieces and dust with powdered sugar.

Applesauce Candy

2 cups applesauce
1-1/2 cups sugar
1/2 teaspoon cinnamon
1/4 teaspoon cloves
2 envelopes unflavored gelatin
1/2 cup cold water
Powdered sugar

Heat applesauce, sugar, and spices, stirring until sugar dissolves. Soak gelatin in water 5 minutes. Add gelatin to apple mixture, stirring well. Pour 1" thick into 8" square buttered pan. Chill until firm. Cut into squares and roll in powdered sugar.

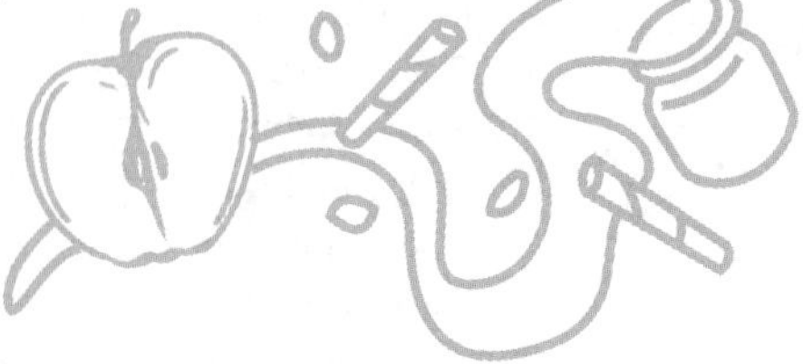

Honey Baked Apple

Apples
Raisins
Honey
Cinnamon

Core apples. Fill each with raisins. Pour honey in center to top of each apple. Sprinkle with cinnamon. Pour 1/2" water in baking dish and bake apples at 375 degrees for 50 to 60 minutes.

Sugar Coated Baked Apples

6 apples
Sugar
1/2 cup sugar
1 egg
2 tablespoons melted butter
1/2 cup flour
1 teaspoon cinnamon
1/4 teaspoon salt

Peel and core apples. Roll in sugar and put in baking dish. Fill centers with 1/2 cup sugar. Beat together remaining ingredients. Drop 1 tablespoon batter into each apple. Bake at 375 degrees for 50 to 60 minutes.

Perfect Baked Apples

6 apples
Brown sugar
Cinnamon
Butter

Core apples and pare a 2" strip along top. Place in baking dish. Fill centers with brown sugar, a sprinkling of cinnamon and 1/2 teaspoon butter. Pour 1 cup water around apples and bake at 375 degrees for 50 to 60 minutes.

Applesauce Fruit Cake

3 cups sweetened applesauce
1 cup shortening
1/2 cup sugar
4-1/2 cups flour
4 teaspoons baking soda
1 teaspoon nutmeg
1/4 teaspoon cinnamon
1-lb. mixed candied fruits
1-lb. raisins
1/4-lb. nuts

Heat applesauce, shortening, and sugar, stirring until sugar dissolves. Cool. Mix in remaining ingredients. Bake at 300 degrees about 2-1/2 hours in greased loaf pans.

Dried Apple Cake

1/2 cup dried apple slices
1/2 cup cider
1/2 cup butter
1 cup sugar
2 eggs
1/2 cup buttermilk
1 teaspoon baking soda
2 cups flour
1/2 teaspoon salt
1/2 teaspoon cinnamon

Simmer apples and cider for 10 minutes. Beat sugar and butter. Add eggs. Mix buttermilk and soda together. Sift flour, salt, and cinnamon. Add flour and buttermilk alternately to creamed mixture. Stir in apples. Bake in greased 9" x 13" pan at 350 degrees for 30 to 35 minutes. Frost with dried apple icing.

Dried Apple Icing

1/2 cup dried apple slices
1/2 cup cider
3-1/2 cups powdered sugar
1/4 cup melted butter

Simmer apples in cider for 10 minutes. Mix sugar and butter. Add apple mixture and beat until creamy.

Custard Apple Cake

2 cups flour
1/2 cup oatmeal
1 cup brown sugar
3/4 cup butter
1/4 teaspoon nutmeg
1 cup water
1 cup sugar
3 teaspoons cornstarch
1 teaspoon vanilla
2 apples

Mix flour, oatmeal, sugar, butter, and nutmeg together. Save 1 cup. Pat rest of mixture on bottom of baking dish. Combine water, sugar, and cornstarch in saucepan over medium heat until thick. Add vanilla.

Slice apples into mixture. Pour into baking dish and sprinkle on crumbly mixture. Bake at 350 degrees for 50 to 60 minutes.

Apple Butter Cake

2-1/2 cups flour
1-1/2 teaspoons baking soda
1 teaspoon cinnamon
1/2 teaspoon nutmeg
1/4 teaspoon salt
1/2 cup butter
1 cup sugar
4 eggs
3/4 cup buttermilk
1-1/2 cups apples
Butter
Powdered sugar

Sift flour, baking soda, spices, and salt. Beat butter and sugar until fluffy; add eggs. Beat in flour mixture alternately with buttermilk. Stir in apple butter.

Pour into greased 9" x 13" baking pan. Bake at 350 degrees for 50 minutes. Cool. Sprinkle with powdered sugar.

Honey Applesauce Cake

1/2 cup shortening
1 cup honey
3 cups flour
1-1/2 teaspoons baking soda
1/2 teaspoon salt
1 teaspoon cinnamon
1 teaspoon nutmeg
1/4 teaspoon cloves
1-1/2 cups applesauce
1/2 cup raisins
1/2 cup chopped walnuts
Powdered sugar

Cream shortening and honey. Sift flour, soda, salt, and spices. Beat into creamed mixture alternately with applesauce. Stir in raisins and nuts. Pour into a greased 9" x 13" pan and bake at 350 degrees for 40 to 50 minutes. Cool and cut into squares. Sprinkle with powdered sugar.

Austrian Apple Torte

1-1/3 cups sifted flour
1/4 cup sugar
1 teaspoon grated lemon rind
1/2 teaspoon salt
1/2 cup butter
1 egg yolk
1 tablespoon lemon juice
2 tablespoons flour
2 tablespoons sugar
1/2 teaspoon cinnamon
3 medium apples
3/4 cup apple jelly

Stir 1-1/3 cup flour, 1/4 cup sugar, lemon rind, salt, and butter in bowl until crumbly. Stir in egg yolk and lemon juice. Shape into ball. Press into bottom and 1-1/2" up sides of greased 9" springform pan. Combine 2 tablespoons flour, 2 tablespoons sugar, and cinnamon. Sprinkle half over dough. Peel, core, and slice apples thinly onto dough. Sprinkle with flour mixture. Heat jelly and spread over apples. Bake at 375 degrees for 45 to 50 minutes. Cool. Loosen from pan.

Apple Pudding Cake

- 1 cup sugar
- 1/2 cup butter
- 1 egg
- 1 teaspoon baking soda
- 1 teaspoon cinnamon
- 1/4 teaspoon nutmeg
- 1/2 cup cold coffee
- 1-1/2 cups flour
- 1/4 teaspoon salt
- 1 cup raisins
- 1/2 cup chopped nuts
- 2 cups chopped apples

Cream sugar and butter. Add egg. Beat in coffee and dry ingredients alternately. Stir in nuts, raisins, and apples. Bake in 9" x 13" pan for 45 minutes at 350 degrees. Cut in squares and pour apple pudding sauce on top.

Apple Pudding Sauce

1/4 cup butter
1/2 cup sugar
1 egg yolk
1-1/2 tablespoons flour
1/4 teaspoon nutmeg
1 teaspoon vanilla
1-1/4 cups water
1 apple, diced

Cream butter and sugar. Beat in egg yolk, flour, nutmeg, and vanilla. Gradually add water. Cook until thick, stirring constantly. Mix in apples. Serve over cake.

Crunchy Caramel Apple Pie

About 80 jillion calories a slice, but who cares?

9" unbaked pie shell
30 caramels
3 tablespoons water
5 large apples
3/4 cup flour
1/2 teaspoon cinnamon
Dash nutmeg
1/4 cup sugar
1/3 cup butter
1/2 cup chopped walnuts

Melt caramels with water in saucepan over low heat, stirring occasionally until melted. Peel and core apples. Slice half the apples into pie shell. Pour half the caramel over apples. Repeat with apples and sauce.

Mix flour, spices, and sugar in small bowl with butter until crumbly. Stir in nuts. Sprinkle over pie. Bake at 375 degrees for 45 minutes.

Crispy Apple Cranberry Pie

Serve with ice cream.

9" unbaked pie shell
5 large apples
1 cup canned whole cranberry sauce
1/2 cup brown sugar
1 tablespoon flour
1/2 teaspoon nutmeg
1/2 teaspoon cinnamon
1/2 cup brown sugar
3/4 cup flour
1/4 cup butter

Peel and core apples. Slice into bowl and mix with cranberry sauce, 1/2 cup brown sugar, 1 tablespoon flour, and spices. Pour into pie shell.

Stir remaining ingredients together in small bowl until crumbly. Sprinkle over pie. Bake at 400 degrees for 35 minutes.

Apple Cider Pie

2-crust, 9" unbaked pie shell
4 cups boiling apple cider
3 cups dried apple slices
3/4 cup brown sugar
2 tablespoons flour
1/2 cup raisins
1/2 teaspoon cinnamon
1 tablespoon lemon juice
3 tablespoons butter

Serve with ice cream.

Pour cider over apples and let stand 1 hour. Drain apples, reserving 1/2 cup liquid. Mix apples, sugar, flour, raisins, and cinnamon and pour into pie shell.

Pour lemon juice and reserved liquid over mixture. Dot with butter. Cover with top crust and bake at 400 degrees for 45 minutes.

Applesauce Custard Pie

9" unbaked pie shell
3 eggs
5 tablespoons sugar
1/2 teaspoon cinnamon
2-1/2 cups milk
1 cup applesauce

Beat eggs, sugar, cinnamon, and milk. Mix in applesauce. Pour into shell. Bake at 450 degrees for 15 minutes. Reduce heat to 350 degrees for 30 minutes.

Apple Butter Pie

9" unbaked pie shell and lattice top
1/2 cup apple butter
2 eggs
1/2 cup sugar
1-1/2 tablespoons cornstarch
2 cups milk

Mix apple butter, eggs, sugar, and cornstarch well. Beat in milk gradually. Pour into pie shell. Top with lattice strips. Bake at 350 degrees for 35 minutes.

Paper Bag Apple Pie

9" unbaked pie shell
5 large apples
1 cup sugar
2 tablespoons flour
2 teaspoons lemon juice
1 teaspoon cinnamon
1/2 teaspoon nutmeg
1/2 cup flour
1/2 cup brown sugar
3 tablespoons butter

Peel and core apples. Slice into large bowl and toss with sugar, 2 tablespoons flour, lemon juice, and spices. Pour into pie shell. Mix remaining ingredients together in small bowl until crumbly. Sprinkle over apples.

Place pie in large, brown paper bag. Fold open end over twice. Staple shut or seal with large paper clips. Bake at 400 degrees for 50 minutes.

Brandy & Raisin Apple Pie

2- crust, 9" unbaked pie shell
1/2 cup raisins
3 tablespoons brandy
5 large apples
1 cup sugar
1 tablespoon cornstarch
1 teaspoon lemon
1 teaspoon cinnamon
1/4 teaspoon nutmeg
3 tablespoons butter

Heat raisins and brandy in saucepan over low heat. Peel and core apples. Slice into bowl and toss with sugar, cornstarch, lemon juice, and spices.

Stir in raisins and brandy. Pour into pie shell and dot with butter. Put on top crust and slit for steam. Bake at 450 degrees for 45 to 50 minutes.

Apple and Cheese Pie

9" unbaked pie shell
5 large apples
2/3 cup sugar
2 tablespoons flour
1 teaspoon cinnamon
1/2 cup sour cream
1/4-lb. cheddar cheese, grated

Peel and core apples. Slice and arrange in pie shell. Mix sugar, flour, cinnamon and sour cream. Pour over apples. Sprinkle cheese on top.

Bake at 400 degrees for 10 minutes. Reduce heat to 375 degrees and bake 40 minutes.

Brown Sugar Apple Pie

2-crust, 9" unbaked pie shell
5 large apples
1 cup brown sugar
2 tablespoons flour
1-1/2 teaspoons cinnamon
1/4 teaspoon nutmeg
Dash mace
1 tablespoon grated lemon peel
2 tablespoons butter

Peel and core apples. Slice thin and mix with sugar, flour, spices, and lemon peel. Fill pie shell with apples. Dot with butter. Put on top crust and slit for steam. Bake at 400 degrees for 50 minutes.

Apple & Cream Pie

9" unbaked pie shell
5 large apples
1/2 cup sugar
1/2 cup water
1/2 teaspoon cinnamon
Dash salt
1 egg
1/2 cup heavy cream

Peel and core apples. Cut into eighths. Bring to a boil in saucepan with sugar, water, and spices. Cover and simmer 15 minutes. Drain, reserving syrup.

Put apples in pie shell. Beat egg and cream together with syrup. Pour over apples. Bake at 350 degrees for 30 minutes.

Deep Dish Apple Pie

Pastry for 8" dish
5 large apples
3 tablespoons quick-cooking tapioca
3/4 cup sugar
1/4 cup brown sugar
Dash salt
1 teaspoon cinnamon
1/2 teaspoon nutmeg

Peel, core, and slice apples into thick pieces. Mix all ingredients together except pastry. Spoon into buttered deep 8" baking dish. Cover with pastry top. Cut in 6 to 7 slits. Bake at 425 degrees for 35 to 40 minutes.

Apple Crumb Pie

9" unbaked pie shell
5 large apples
1 cup sugar
1 teaspoon cinnamon
1/2 teaspoon nutmeg
1/3 cup butter
3/4 cup flour

Peel, core, and slice apples into pieces. Mix half the sugar with spices and sprinkle over apples. Put apples in pie shell. Mix remaining ingredients together until crumbly.

Sprinkle over apples. Bake at 425 degrees for 10 minutes. Reduce heat to 350 degrees and bake 35 to 40 minutes.

Red Cinnamon Apple Pie

2-crust, 9" unbaked pie shell
5 large apples
1 cup sugar
2 tablespoons flour
3 tablespoons red cinnamon candies
1/4 teaspoon salt
2 tablespoons butter

Peel and core apples. Slice into bowl with sugar, flour, candies, and salt. Mix well. Fill pie shell with mixture. Dot with butter. Put on top of crust. Slit for steam. Bake at 400 degrees for 50 minutes.

Plain Old Apple Pie

2-crust, 9" unbaked pie shell
5 large apples
3/4 cup sugar
2 tablespoons flour
Dash nutmeg
1 teaspoon cinnamon
1/8 teaspoon salt
2 tablespoons butter

Peel and core apples. Slice thin and combine with sugar, flour, spices, and salt. Fill pie shell with apple mixture and dot with butter. Put on top crust and slit for steam. Bake at 400 degrees for 50 minutes.

Dried Apple Pie

2-crust 9" unbaked pie shell
2 cups dried apple slices
2-1/2 cups water
1 orange, rind and juice
3/4 cup sugar
1/2 teaspoon cinnamon

Soak apples overnight in water. Add orange rind and juice and boil until soft, about 30 minutes. Grind in food mill. Mix in sugar and cinnamon. Pour into pie shell and cover with top crust. Cut in slits. Bake at 425 degrees for 10 minutes. Reduce heat to 350 degrees and bake for 30 minutes.

Sour Cream Apple Pie

- 9" unbaked pie shell
- 3/4 cup sugar
- 1 teaspoon vanilla
- 2 tablespoons flour
- 1 cup sour cream
- 1 egg
- 4 apples
- 1/2 cup brown sugar
- 1/3 cup flour
- 1/4 cup butter

Beat together sugar, vanilla, flour, sour cream, and egg. Dice the apples and add to mixture. Pour into pie shell and bake at 400 degrees for 25 minutes.

Mix remaining ingredients until crumbly and sprinkle over pie. Bake for 20 more minutes.